THE ORIGIN OF THE UNIVERSE

A Short Version

By Forester de Santos

The Origin of the Universe / A short version
Forester de Santos

© 2017 Forester de Santos

ISBN-13: 978-1718077645

ISBN-10: 1718077645

All Rights Reserved

No part of this book may be copied, sold or distributed, in either printed or electronic format, without the written permission of Forester de Santos. For more information, please contact Forester de Santos at Forastero21amado@gmail.com.

Kindle Edition

The Origin of the Universe / A short version
Forester de Santos

Prolog

Truly blessed is that man that has true knowledge of how existence or of how the universe truly functions because that man also knows how that man himself truly functions.

And he also will do as existence her very self to continue on existing or living on but existing or living on as if forever new and in complete abundance…

The Origin of the Universe / A short version
Forester de Santos

Acknowledgement

I would like to give thanks to my son Fred for suggesting me to write this short Book version of the origin of the universe.

I would like also to give Fred thanks for helping with the drawings in this Book.

Thanks so very much, Fred!!

The Origin of the Universe / A short version
Forester de Santos

Dedication

This short version of the origin of the universe is dedicated to the good reader and to those who joyfully fight or joyfully struggle or contend to understand the truth and thus enter in the truth as the truth herself.

The Origin of the Universe / A short version
Forester de Santos

Table of Contents

Prolog...pg.3
Acknowledgement...pg. 4
Dedication...pg. 5

Introduction - it can be done!...pg. 7
Chapter One - what makes the universe?...pg. 11
Chapter Two - many numbers of universes...pg20
Chapter Three - nothing is something!...pg. 26
The Conclusion - as we have proven!...pg. 29

About the author...pg. 31
Additional Notes...pg. 33
Who am I really?...pg. 64

The Origin of the Universe / A short version
Forester de Santos

Introduction

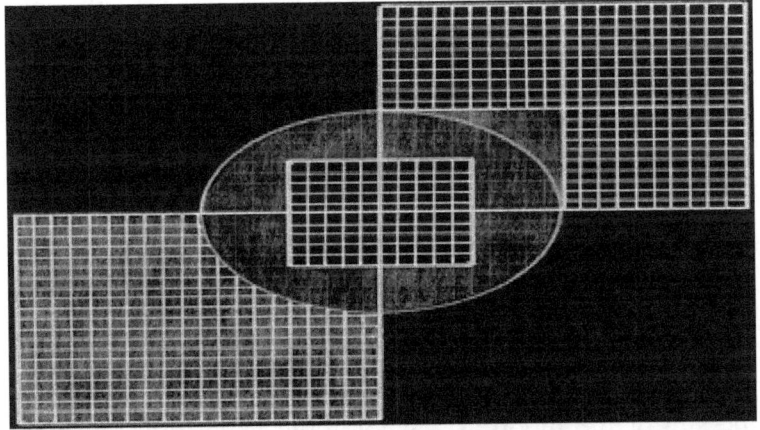

Explaining or understanding the true or the only origin of the universe or the only origin of all of existence is not a very easy task, but it can really be done!

Explaining or understanding the true origin of the universe or existence requires both a simple scientific theory and a simple scientific fact or a simple physical fact that can be truly or physically proven.

Explaining the true origin of the universe or existence requires really describing something physical that we

The Origin of the Universe / A short version
Forester de Santos

cannot really see (with some of the physical or the solid pieces or physical facts that we already have).

We cannot see our face in front of us unless there is a physical reflection or a there is a physical sensation of some kind.

And it is easier to theorize or fantasize than it is to believe or see reality.

Our natural or our animal instinct is to see in parts or see in pieces, not whole and but only when those parts or those pieces are in motion.

We cannot see the whole picture unless we see parts of it first. It is also an animal instinct to see in parts or pieces and not see whole.

An animal, for example, cannot see a tree. An animal can only see parts of that tree and even though that animal can see different parts of a tree, that animal cannot see the whole tree. Animals through instincts can only see light and shades and the motion that goes between.

Again, the animal instinct or the animal mind cannot see the sun or the moon as a whole object, but only the light and the shade and the physical motion that is done in them.

To describe something solid, therefore, we then have to either take it apart or put it together from the solid pieces we have at hand.

From the missing solid piece or pieces we then theorize, thus making reality or the fact at hand difficult to see or to understand.

The theory is the simple glue to simply glue the pieces or reality or the fact that we already have at hand, but too often the glue is more than reality and thus makes more

The Origin of the Universe / A short version
Forester de Santos

puzzles or more pieces than those that we already have at hand.

And so most of us become more interested in the very simple but very messy glue, filling countless books having nothing to do with the reality or the solid pieces we started with.

What irony and what a waste!

Even as well trained scientists, we are still prone to our own prejudices, beliefs or the lack of physical or solid information.

For example: Which of the two following questions below has a possible or a true answer?

"Why is the universe round?"

Or "What makes the Universe round?"

The "Why" question is a very childish question!

It is asked as if no answer is really wanted or really expected from the person asking the childish or the foolish question.

At the very same time, that childish question keeps the person from seeking an answer, any answer!

And if the childish question were asked to another person, that other person too would be stuck no matter how smart he was and he would not even attempt to give an answer.

Remarkably, that very childish or very foolish question is asked by most scientists and professionals than it is really asked by children!

The Origin of the Universe / A short version
Forester de Santos

On the other hand, "What makes the universe round?" has an answer.

It is a very simple question that although will lead to many more simple questions, it will give one final and one very simple answer!

The Origin of the Universe / A short version
Forester de Santos

Chapter One

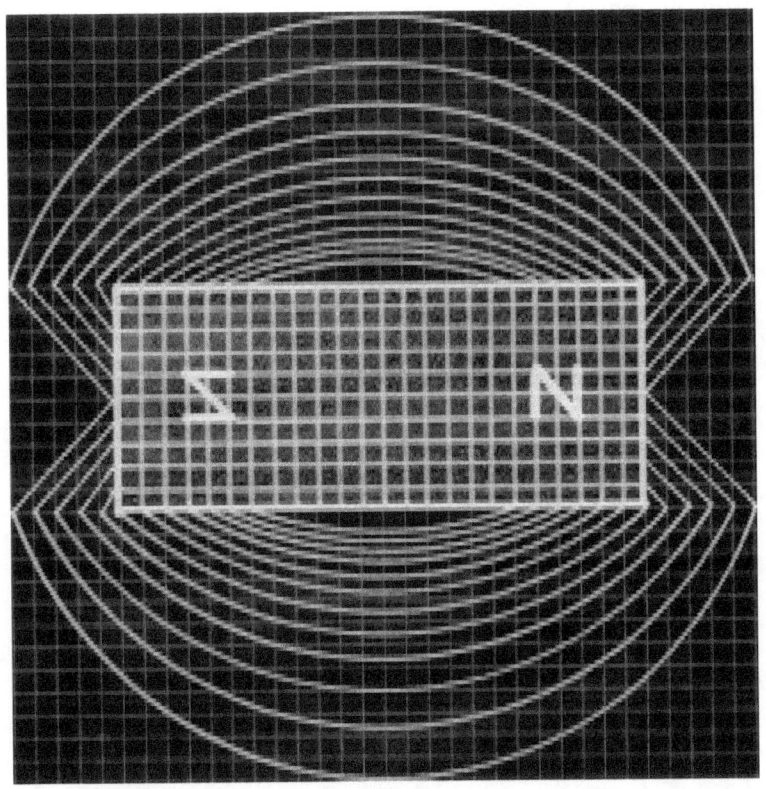

What makes the universe round?

The Origin of the Universe / A short version
Forester de Santos

To answer that very simple question, one has to also ask: What makes the universe?

That is to say in truth, what is the universe made of?

The universe is made up of space, darkness, and cold or coldness.

The universe is also made up of matter, light, and of heat.

The universe is also round or sphere-like and the universe is also in motion.

What is space?

Space is really the simple absence of matter. Space has no mass.

What is darkness?

Darkness is also really the simple absence of light. Darkness has no mass.

And what is cold or coldness?

Cold or coldness is also really the simple absence of heat. Cold or coldness has no mass.

What is matter?

Matter is really that which simply occupies space and has mass.

What is light?

Light is really that which simply occupies space and has mass.

What is heat?

The Origin of the Universe / A short version
Forester de Santos

Heat is really that which simply occupies space and has mass.

Anything that exists, no matter how light has mass. Even light has mass. Light bends as light reaches or passes by a gravitational field or magnetic pole.

And if light being also heat bends, then heat or light has mass.

Now let's say that space, darkness and cold or coldness is the same thing.

And let's say also that matter, light and heat are the same thing.

Now, we have two opposites to work with: Space and matter, Non-existence and Existence, Negative and Positive, Black and White.

Space and matter form the simple universe which is round or sphere-like where space and matter unite or intersect or even separate.

What two simple figures or what two simple objects make a sphere or make a circle where they physically unite or physically join or even divide?

What about two magnets?

What about two magnets in the simple form of two cubes?

Yes!

Where the two magnetic cubes, in this case matter and space, physically meet or join or unite, at the corners, they actually form a sphere.

The Origin of the Universe / A short version
Forester de Santos

In fact, many spheres or circles are formed by the two cubes or the two magnets.

But how are the spheres or the circles formed?

The spheres or the circles are formed due to the magnetic fields of each of the magnets or cubes.

An excellent and also a very simple example is the field of a bar magnet.

See figure number one below, the bar magnet.

Figure Number One: The Bar Magnet

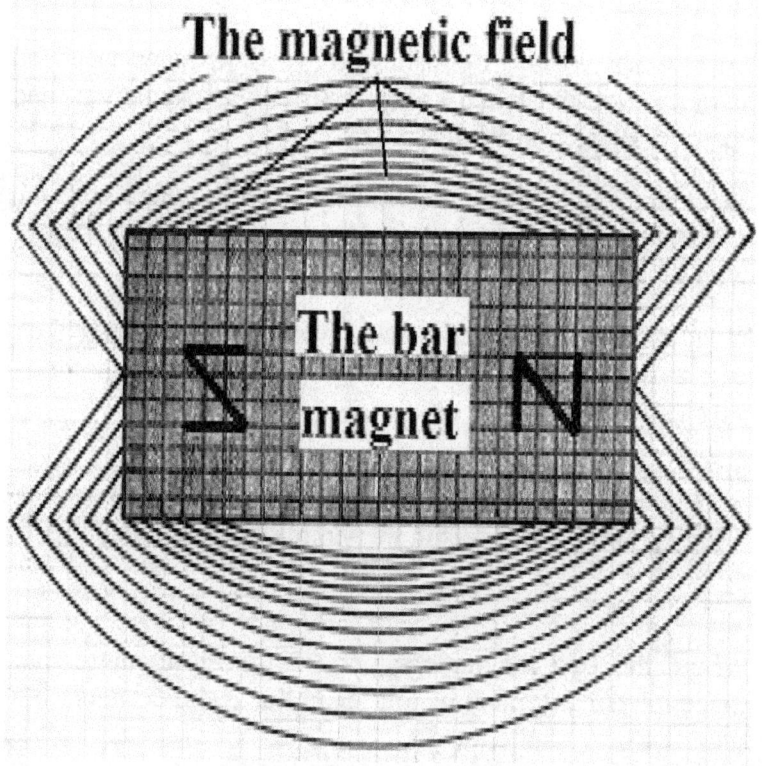

The Origin of the Universe / A short version
Forester de Santos

The magnet creates a special physical condition in the space around the magnet.

That space or condition is noted as a magnetic field.

The cubes' attraction for each other, therefore, makes possible the spheres.

That is to really say, instead of the magnetic field going around each of the cubes as it does with a bar magnet, the field turns toward the opposite cube, thus making the possible spheres or the circles as in figure number two below.

Figure Number Two: The Magnetic Spheres

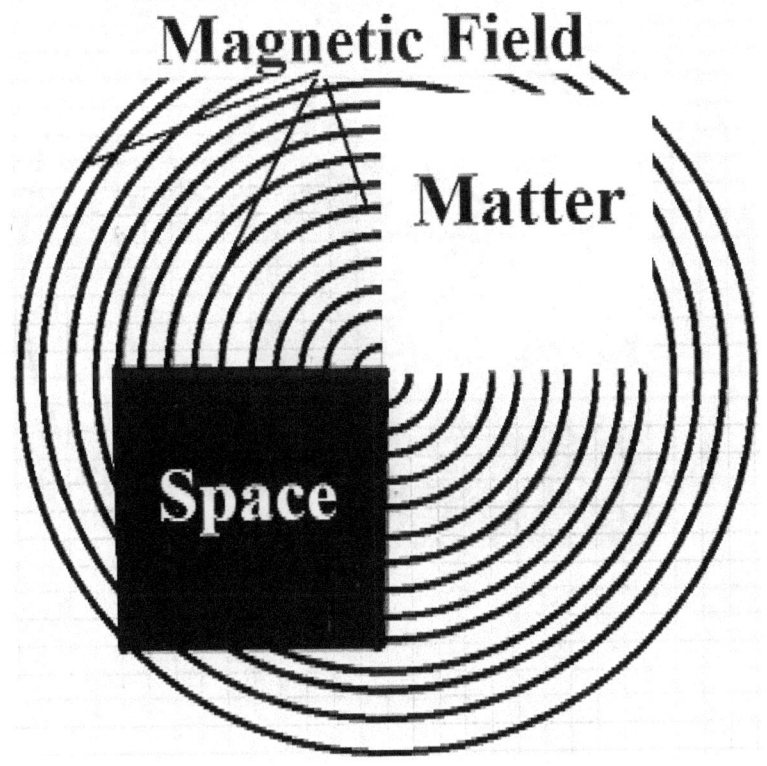

The Origin of the Universe / A short version
Forester de Santos

Now, inside one of those spheres or circles is the universe!

When the two magnets, in this case space and matter, attract one another through their magnetic fields, the two magnets create magnetic spheres.

Inside one of those magnetic fields or spheres is our simple universe.

See figure number three below, the universe.

Figure Number Three: The Universe

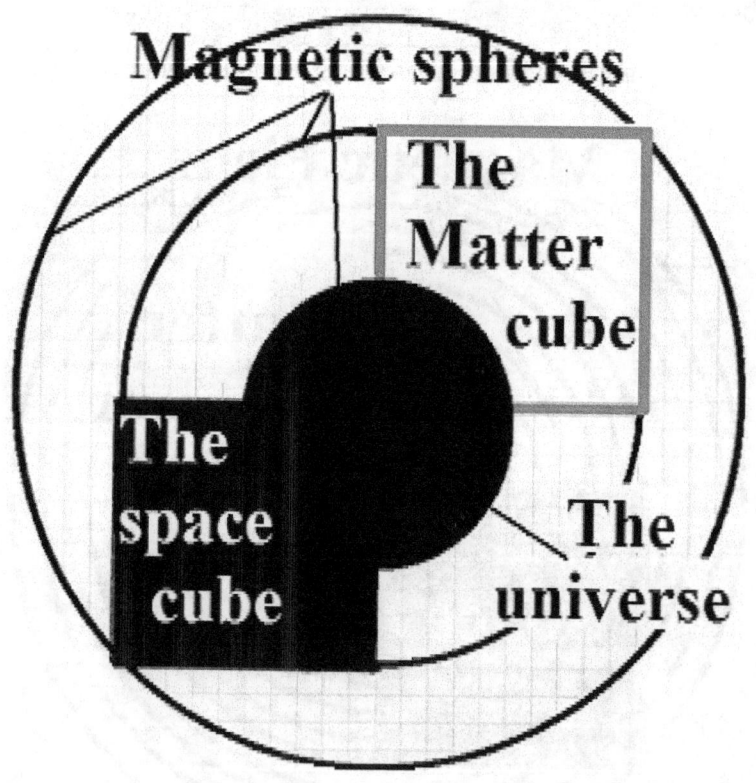

The Origin of the Universe / A short version
Forester de Santos

But what makes possible the cube or cubes?

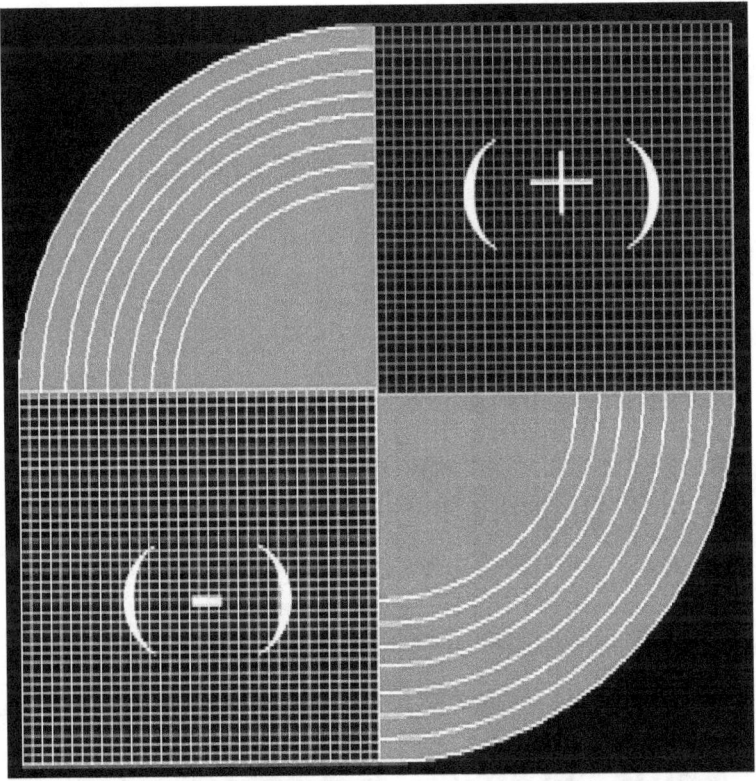

Since the cube is outside of space, there is no compression or vacuum of any kind to act or react upon the cube or cubes, allowing the cube or the cubes to expand throughout and, therefore, making or taking its or their shape.

An interesting thing about empty space: space is a great imitator and space allows for the magnetic field of matter to go through and thus space acts like a magnetic.

We now have two magnetic cubes and where they meet, countless of spheres or circles are formed; thus the universe.

The Origin of the Universe / A short version
Forester de Santos

One of those spheres or circles is the universe or one universe.

Actually, every sphere or every circle is a separate universe!

Both cubes attract one another with the same magnetic force.

That is to say, the cubes attract one another with equal magnetic force or magnetic attraction.

The matter cubes attracts the space cube, but the space cube releases nothing for space has nothing to release.

The space cubes attracts the matter cube, which has density, and a part of the matter cube is released and takes up space in the sphere or the circle formed by the two cubes.

See figure number four below, matter in space.

Figure Number Four: Matter in the Sphere

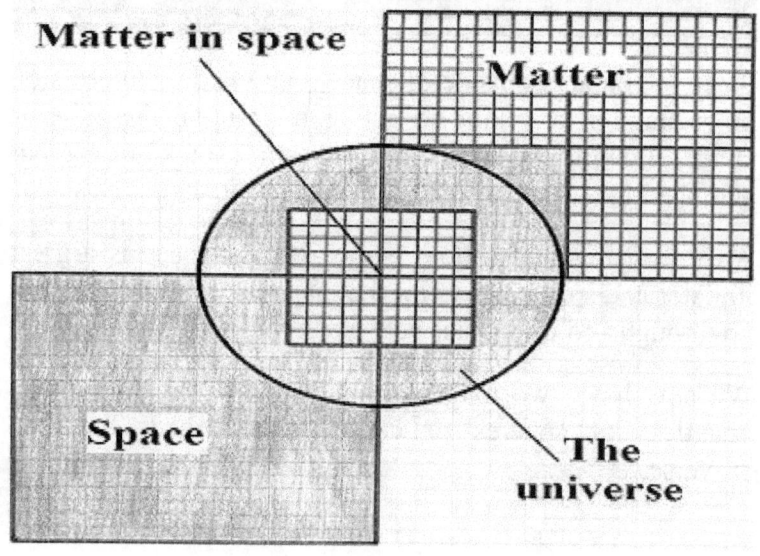

The Origin of the Universe / A short version
Forester de Santos

Once the physical part or the physical piece from the infinite matter cube, which is in form of a smaller solid cube, enters the vacuum of space or enters one of the magnetic spheres, the physical properties of matter in the vacuum of space change.

Matter in the vacuum of space shrinks and then expands throughout the universe or throughout the sphere or the circle created by the two original cubes, the space cube and the matter cube.

Thus, the big bang!

See figure number five below, the expanding universe.

Figure Number Five: The Expanding Universe

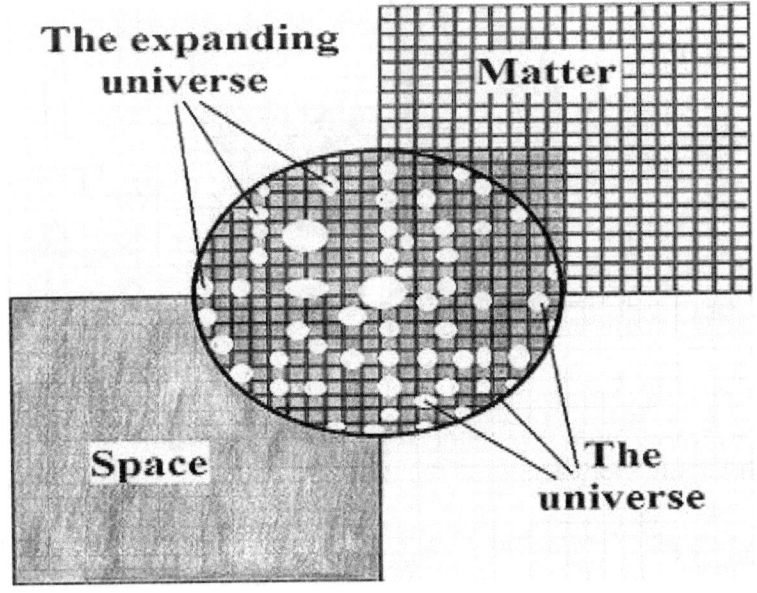

Chapter Two

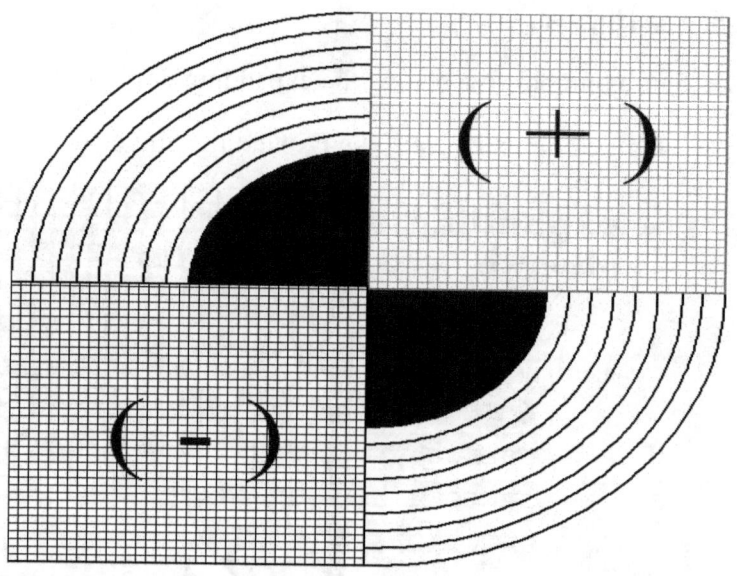

In reality, many numbers of universes or dimensions of time are between the two infinite magnets: space and matter.

Every sphere or every circle in figure number two illustrates a separate universe or a separate dimension of time.

The Origin of the Universe / A short version
Forester de Santos

In figure number two, for example, there are fourteen spheres or fourteen circles representing fourteen spheres or fourteen circles.

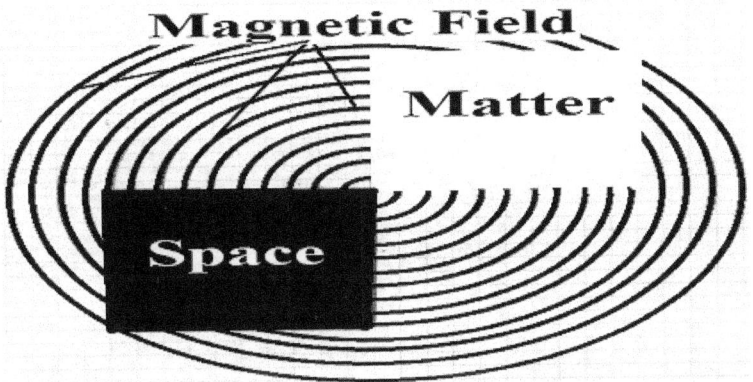

However, even though the space cube and the matter cube expand forever and the spheres or the circles are limitless as are the numbers are limitless, the numbers of universes or the number of dimensions of time are limited.

What we have is a sphere inside a sphere inside a sphere inside a sphere all the way until infinity.

Every sphere, every circle or every universe is separated by a dimension of time or separated by a dimension of space.

The circles or the spheres or the universes as well as the dimensions of time increase in size as the two cubes extend forever and ever.

That is to say in truth, every sphere or every universe increases in size by one. Sphere number two or universe number two, for example, is twice the size of sphere number one or universe number one.

Sphere number one or universe number one is three times smaller than sphere or universe number three.

The Origin of the Universe / A short version
Forester de Santos

The motion or the expansion in time or in space is created by the simple magnetic field of the two infinite cubes and the big bang explosion.

See figure number six below, space-time.

Figure Number Six: Space-time

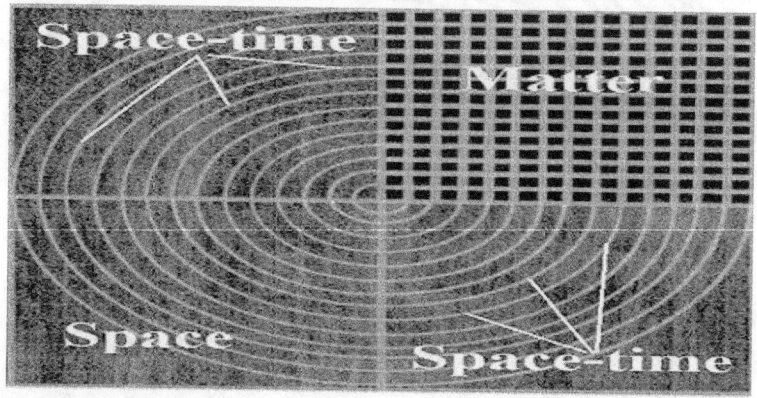

In figure number six above, notice that there are now two more cubes. The two extra cubes are the time cubes or the space time-cubes.

Although those two extra cubes are not directly responsible for time, most of the magnetic fields or spheres and the expansion of the universe or universes are within those two extra cubes.

The two extra cubes also indirectly add more space to the universe or universes. Henceforth, the cubes are noted as space-time.

Every sphere is a dimension of time and a separate universe. The dimensions of time are finite in numbers as the universes are also finite in numbers.

When the space cube attracts the matter cube, not only does the matter cubes releases a smaller cube, but also the matter

The Origin of the Universe / A short version
Forester de Santos

cubes releases multiple smaller matter cubes at the very same time.

The matter cube releases smaller matter cubes according to the size of the spheres or the universes.

That is, the matter cube releases at the very same time one matter cube for the number one sphere or universe; the matter cube releases two matter cubes for the number two sphere or the number two universe; the matter cube also releases three matter cubes for the number three sphere or for the number three universe and so on.

In the case of figure number six, fourteen different universes or dimensions of time began spontaneously.

As soon as a solid piece of matter from the giant or from infinite matter cube enters the vacuum of space or the magnetic spheres created by the magnetic field of both the giant or infinite matter cube and the giant or infinite space cube, matter in the vacuum of space or in the magnetic spheres or in the magnetic rings is super compressed due to the enormous pressure or the enormous vacuum of space.

The enormous pressure causes the smaller solid matter cube in the vacuum of space to compress into a giant sphere of solid matter and due to the compression and superheat thus making possible the big bang!

Or the big bangs!

Enormous amount of energy is loss because of the big bangs, but a lot matter is still left in the vacuum or compression of space or the magnetic sphere to expand throughout space and the universe or universes.

Some matter, like light or less heavy, will lose energy, therefore, turning to or becoming many planets and moons, meteors or comets and some matter becoming even smaller pieces, such as space dust.

The heaviest or the largest matter or the heaviest of stars or suns will turn or will transform into supernovas, really collapsed stars and then turn into giant black holes.

There will be another point in time, millions of years in the future, when all of the universe and all the universes or all the dimensions of time will be occupied by black holes or dead stars.

The universe and universes will look like a Swiss cheese, but without the sweet smell.

See figure number seven below.

Figure Number Seven: The Imploding Universe

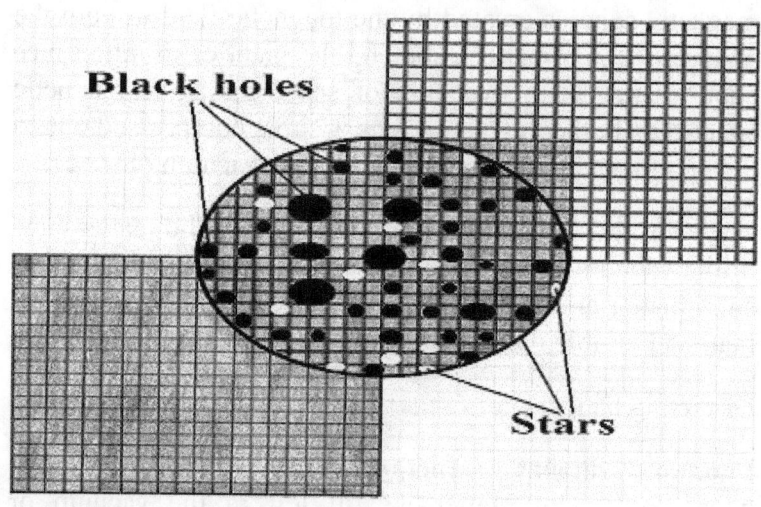

At that stage, when our entire universe and all the other universes or all dimensions of time are occupied by many giant black holes, the giant black holes will disintegrate or take apart all matter in our universe and all the other universes spontaneously.

The Origin of the Universe / A short version
Forester de Santos

That is, the giant black holes will begin contracting the universe or universes until our entire universe and all other universes all collapse--the opposite of the big bangs!

The black holes will begin to move to the center of the collapsed universe and universes, thus making one super giant or enormous black hole or one super giant or enormous vacuum per universe.

However, the super black holes increase in size as the universe increase in size.

See figure number eight below, the super-giant black hole.

Figure Number Eight: The super-giant black hole.

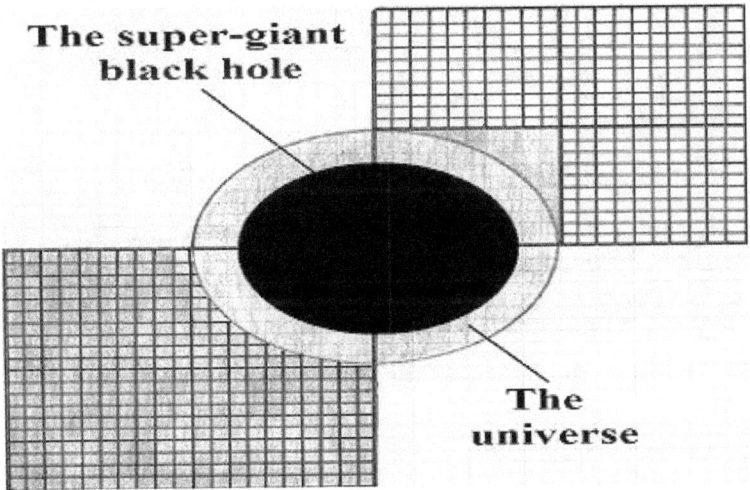

It is this super-giant or enormous black hole vacuum that attracts new matter from the matter cube.

Once new matter enters the super-giant black hole vacuum, matter is super compressed thus recreating the big bangs and destroying the super-giant black hole vacuum!

Chapter Three

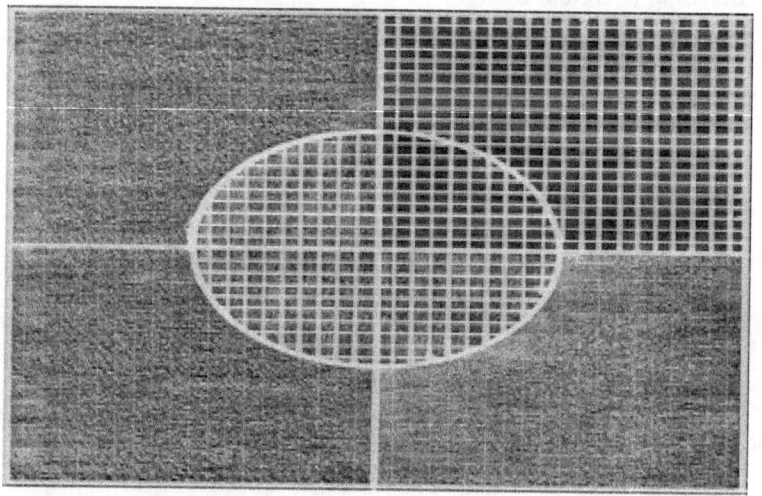

Existence or the universe always was and will always be.

Nothing is something.

Nothing or nothingness or emptiness is part of existence or the universe.

Nothing or non-existence exists and thus also makes possible existence or the universe.

Non-existence or space happens by itself and thus also makes possible existence or the universe.

The Origin of the Universe / A short version
Forester de Santos

The universe or existence is a product of something, matter; and nothing, space.

The interaction of something with nothing allows for the universe to be possible or to exist.

The universe and time are products of Non-existence and existence.

That is to say in truth, the universe and time are products of two opposites: space and matter, negative and positive, non-existence and existence, vacuum and expansion, black and white.

Where those two opposites meet or intersect, there is the universe and time.

Time, which is motion or expansion, is created by the magnetic fields or the magnetic attraction of the two opposites: space and matter.

The space cube and the matter cube or simply all of existence is really a very simple electric motor, really a simple magnetic motor or generator.

The electric or the magnetic motor is made possible by a very simple magnetic field.

In other words, the magnetic field makes possible the simple motion in the motor.

In the same way, the magnetic field of the two infinite cubes makes possible motion and thus time.

Time, like the simple universe and simple existence, always was and will always be.

Time, which is motion or expansion, had no beginning and will have no end.

The Origin of the Universe / A short version
Forester de Santos

The motion or expansion in time is created by the magnetic field of the two infinite cubes.

Time is also a point of view.

The Conclusion

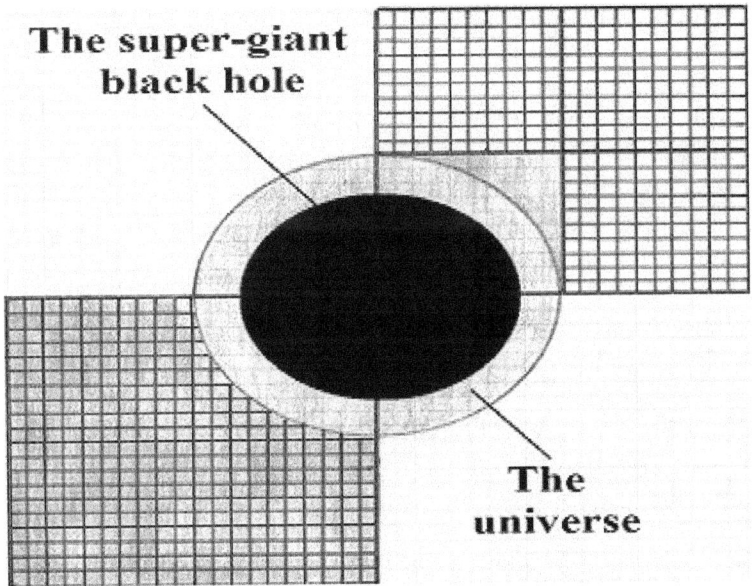

As we have seen and as we have proven, the universe, time or simply existence is a simple product of two infinite magnets in the form of cubes: space and matter.

By using the Periodic table of the Elements, which is a simple diagram of a bar magnet or electric motor or generator, we have proven the above statement.

And also, by using the structure of the atom, which is also a simple microscopic diagram of a simple bar magnet or

The Origin of the Universe / A short version
Forester de Santos

electric motor or generator, we have also proven the first statement of this conclusion.

And in conclusion, the most important thing in existence or in the universe is conscious thought!

Conscious thought only exist in conscious living beings. Conscious thought is not an automatic process as existence is.

Conscious thought is not a natural process and conscious thought really requires effort from the conscious living being.

Even though knowledge simply exists, the conscious living being does not know or does not learn until he makes an effort to think and ask that unnatural question which will bring him knowledge or recollection, which is in truth recognition or acknowledgement!

The only beginning in time and existence or the universe is really conscious thought.

The only limits that the conscious living being has are the limits that the conscious living being gives or sets him himself.

So, let's think or change our beliefs or points of view!

One day, hopefully very, very soon, thinking or conscious thought will be the most important commodity or asset a living being will possess or have.

Thinking or conscious thought or conscious effort will one day keep the universes as well as the living conscious beings from collapsing!

The Origin of the Universe / A short version
Forester de Santos

About the author

I truly hope that you enjoyed reading this short work as much as I enjoyed writing it.

I was born in the Caribbean Island of Puerto Rico, where I am living with my youngest of two sons.

As a child and most of my adulthood, I lived in the States. I mostly lived in the City of New York, mostly in the South Bronx, sometimes called Fort Apache.

I went to Marist College in the town of Poughkeepsie, Up State New York; and later transferred to Hunter College, in the City of New York. In 2008, I moved here back to Puerto Rico.

I am one of those, and perhaps the only one in these very difficult times, who with complete sincerity seek the truth the truth of the universe, the truth of the creation of God and the truth of what is the salvation of man and man, as far as I know, will not be saved from eternal death by the technology of man as neither the very sword will save man.

The Origin of the Universe / A short version
Forester de Santos

Technology as the sword will only save man but save man only temporarily. And technology can also be a double edge sword or contra productive because man does not really use it to expand or to increase his time or to expand or to increase his real knowledge in truth.

My search for the truth of the universe, for the truth of the creation of God and for the truth of what is really the salvation of man, has done to me very good, so good in fact that it has taken me to not only achieve or to have real faith of God, but also my search has allowed me to really understand what is in truth the salvation of man by God the good and loving Creator.

And the very simple truth is that without real knowledge of God, thus there will not be the salvation of God and without the salvation of God, thus there will neither be the salvation of man nor will there be the salvation of the universe, because as man will return eternally dead to the very dust, thus also the universe will return to nothing...

Thus, in my very simple writings there are certain truths or real knowledge of life!

Glad and happy he who in truth lends attention!

The Origin of the Universe / A short version
Forester de Santos

Additional Notes

The Stars and the Black Holes

The universe is a neutral point or is a point zero which is converted into a positive point or into a one or more when light or matter enters into the universe.

The universe also is converted into a negative point when the grand majority of light turns off or the majority of matter no longer has energy...

Before light or matter enters into the universe the weight of the universe is cero and when light or matter enters into the universe thus the universe takes on weight.

Light or matter is composed of 118 Elements and the weight of each element is two times its atomic number, more like its positive number.

In the case of element number one, for example, its weight is of two and in the case of element number two its weight is of four and in the case of element number 118 its weight is of 236...

Curiously, that the totality of the 118 elements adds to one and that the totality of their weight adds to 2.

That is to say, that if we added from one to 118 thus the sum would be 7,021. And if we added that sum thus it

The Origin of the Universe / A short version
Forester de Santos

would be 10 and if we add that last sum thus it would be one.

That is, 118 is equal to one...

And if we added from two to 236 the sum would be 14, 042. And if we added that sum thus it would be 11 and if we added that last sum thus it would be 2.

That is, 236 are equal to two...

Thus, light or matter enters into the vacuum of space or into the universe as one or as a unit which is composed of 118 pieces or elements and the weight of the element is two times the atomic number of the element.

Thus in truth, number one itself is composed of 100 percent plus 18!

That is to say, that the number one or even oneself is equal to 118 percent!

Also light or matter could enter into the vacuum of space or into the universe with only or as one element with its weight of two but that element would be able to convert into the other elements, even to the element 118 and its double weight of 236...

Element number one, for example, enters into the vacuum of space with its weight of two and it has 117 other possibilities of converting into the other 117 elements.

That is to say, element one is composed of 118 parts or pieces or the 118 percent and element number has 117 possibilities of converting into the other 117 elements according to the weight which element number one maintains.

In the same manner, element number two with its weight of four has 116 possibilities of converting into the other 116 elements or until the element number 118 with its weight of 236...

The Origin of the Universe / A short version
Forester de Santos

When light or an element enters into the vacuum of space, light or the element enters as if it were a piece of magnet or as if it were a bar magnet.

In the vacuum of space the magnet or light or the element is super compressed not only until it takes the form of a sphere or round but also light or the element or the magnet is super compressed until it gets to a very high level of temperature.

And when the temperature gets to its highest level thus light or the magnet or the element super explodes causing the light or the magnet or the element to divide into two parts and the part with less weight, such as the negative part, takes position in the magnetic field and that magnetic field now is a negative field...

The superior or the positive or the heaviest part of light or of the magnet or of the element takes position in the center or in the nucleus.

Thus, now we have the negative part of the magnet going around the positive part when before they were united and the center or the nucleus was neutral...

And even though the element lost weight because of the explosion or because of bursting, the element continues the one for two.

That is, its weight continues of two although the element one now is 0.999 and its weight is double, 1.998...

Thus, now element number one was reduced to about 0.999 with its new weight of 1.998 but to element number one also remains a neutron or even more than one or a neutral part which can be converted or can be transformed into a positive part and that way not only adding the number of the element but also adding its weight even though it will have only one negative particle going around the center or the nucleus or the positive side...

The Origin of the Universe / A short version
Forester de Santos

But if one or light or the magnet or the element does not convert into the next number or into the next element thus it loses its energy and will only be a piece of dead matter in the vacuum of space and it will be removed one day by the black holes...

Thus, light or the element is the same as a negative particle, is the same as a neutral particle and is the same as a positive particle.

In a way, one is equal to a negative portion plus a neutral portion plus a positive portion which totality is of 0.999 after entering the vacuum of space; but in the vacuum of space light or the element is positive even though the vacuum of space is neutral but reacts as if negative because of the vacuum.

And when the element increases its positive part by converting the neutral part into a positive part, the element cannot attract the negative part because of the vacuum of space because now the negative part becomes as if more or its weight increases because of the weight that it receives indirectly from the vacuum of space.

And if there were not positive attracting the negative thus the negative would expand through the vacuum of space and it would stop being negative and it would be dead matter...

Now, an element, in this case a star, which number is high as the same as its weight thus lasts or remains longer in the vacuum of space or in the universe.

But the element or the star becomes heavier while the energy or matter to continue on lasts or it begins to transform from positive to neutral and once neutral, the element or the star practically becomes negative when its excess or super weight attracts the electrons toward the center and thus causing an implosion in where the element

The Origin of the Universe / A short version
Forester de Santos

or the star becomes a nova or a new star but without light or without energy and that way causing an enormous hole in the vacuum of space when before the element or the star occupied the vacuum of space as an element or as a star…

In other words, matter or the star in the vacuum of space changes from positive to neutral and then from neutral to negative.

Thus, + 0 -, in where the negative is a black hole or a super vacuum cleaner in the vacuum of space.

This black hole practically eats or sucks all the matter around it to take all matter out from the vacuum of space or from the universe to make new space for new matter or for another beginning…

But as long as there are black holes in the vacuum of space or in the universe, thus the vacuum of space or the universe is negative and as long as the vacuum of space or the universe continues as or is negative thus it keeps being for something and not for neutral and from neutral or from zero to positive…

Thus, so that the universe becomes neutral or to zero and from neutral or from zero to positive thus all the black holes or super space vacuums must stop from functioning and once the black holes or the super space vacuums stop from functioning for lack of matter or for lack of energy thus the vacuum of space or the universe will return to neutral or to zero…

Now then, this new vacuum in space or in the universe is neutral and has zero energy but even so attracts new matter or new stars or attracts the new or the next beginning which is outside of the universe or outside the vacuum of space…

And once new matter or new stars enter into the vacuum of space or into the universe thus the vacuum or the universe will be positive or one or more…

The Origin of the Universe / A short version
Forester de Santos

Thus, the cycle or the model or the rhythm of (+ 0 -) continues until the end of all the times...

The Origin of the Universe / A short version
Forester de Santos

Creation and the Number

Creation truly is about giving name or naming to start or to begin and giving rename or renaming to continue without beginning again even though to give rename or renaming is to become as if forever new and as if never there were a start or a beginning and neither end…

All of existence truly is composed of numbers or be it positive numbers or be it neutral numbers or be it negative numbers or be it a combination of all the numbers at the very same time, but there will always be a number that is greater than the other negative numbers until that positive number is converted into a neutral number and after that positive number is converted into a neutral number thus also it will be converted with time into a negative number.

In other words, existence really is equal to (+ 0 -), but existence does not return to negative but rather a part of her and that part of her is the universe or matter or light, which returns to or truly is converted from positive to neutral or to zero and much later it is converted to negative or into nothing.

The number zero is not negative or nothing.

The number zero only is a neutral number which could be converted into a positive or negative number.

Existence as matter or as light is composed of three parts which add to 118 percent, 118 times 3 because they truly

The Origin of the Universe / A short version
Forester de Santos

are 118 positive parts, 118 neutral parts and 118 negative parts or (+ 0 -).

Now then, the neutral part or part zero is the universe and it is where it is added, it is subtracted, it is multiplied and it is divided at the very same time.

The positive part is the physical part of existence or the part from where comes out matter or light or the elements which truly are pure matter, but this part of existence appear to be minor than the other parts of existence.

The negative part of existence is the lack or is the vacuum of space or darkness which is composed of three parts, one is space or emptiness and the other two are space-time, which truly are created by the interaction of the magnetic field between the positive and the negative.

When matter from the physical side of existence enters into the vacuum of space, matter even though a single piece, matter enters with its 118 parts assimilating the 118 elements.

That is, if only one element enters into the vacuum of space that element could assimilate or even can come to be converted into the other 117 elements...

But in reality matter enters from the physical side of existence into the vacuum of space in 118 pieces into 118 parts of the vacuum of space or into 118 dimensions at the very same time.

That is to say, into the vacuum of space there enter 118 elements and every element has 118 parts and at the very same time there enter 118 elements into 118 dimensions.

But that does not remain like that, because with every element the 118 parts are multiplied.

In other words, element number one has 118 parts but element number two has 236 parts and element number

The Origin of the Universe / A short version
Forester de Santos

three has 354 parts and this pattern continues on until the last element, which is element number118.

Also, the weight of the element is twice its number.

That is to say, element number one has a weight of two and element number two has a weight of four and element number three has a weight of six.

This pattern also continues on until element number 118.

Interestingly, that all the parts of an element add to the number of that element...

Thus, element number one has 118 parts and if we added those 118 parts thus we would get one.

That is, if we added 1+1+8 thus they would give us10 and if we added 1+0 thus we would get 1.

In the case of element number two, if we added 2+3+6 thus we would get 11 and if we add 1+1 we would get 2.

This pattern continues on until element number 118. Element number 118 has 118 parts and every part also has 118 parts thus giving a sum of 13,924, also adding to one.

Now then, when element number one enters into the vacuum of space element number one enters with one proton, with one neutron and with one electron, (1+0-1). And also element number one enters with a weight of two.

But once in the vacuum of space, space comprises the element and the element is comprised causing the element to super heat and thus also causing the element to burst or explode.

In that burst or explosion the element loses a neutron or a third part and no longer the weight is of two but is less, such as 1.67.

Also the electron was separated and now is spinning around the proton or the nucleus of the element, (+) -.

The Origin of the Universe / A short version
Forester de Santos

In the case in where an element has a high number thus that element will neutrons in its nucleus, such as (+0+) -, -. Or (+0+0+) -, -, -.

This new transformation of the element, in where the electron is separated from the nucleus makes it possible for the element to be united to other elements and that way converting into a mixture of elements or isotopes…

Furthermore, before an element enters into the vacuum of space its three main parts have the same size or weight, ((+) (0) (-).

But once that element enters into the vacuum of space thus that elements is divided into two parts, the nucleus in where new is the proton (+) and the neutron (0), and the outside part in where now is found the electron (-) going around the nucleus, ((+) (0)) (-).

And while the nucleus keeps its size or weight, the electron loses its size or weight because of the interaction or friction which it has with the nucleus or with the center of the element.

The interaction or friction which the electron has with the nucleus or with the center of the element also causes the electron to last less or lasts less time in the vacuum of space.

Once the electron is fused or is exhausted, the element or the nucleus is converted into a neutral element or without energy even though the nucleus is still positive or with protons and neutrons.

But the outside part or the electric field of the element is now a neutral field or the electrons have been converted into neutrons because of lack energy.

In other words, the seven electron rings of the element or the atom now are neutral.

The Origin of the Universe / A short version
Forester de Santos

And just as the element functions thus that way also functions the number and existence herself, but the number or the symbol of the number is only an illustration of the numbers but truly does not show how is the number in existence or outside the vacuum of space in where there is no friction or movement even though there is a magnetic field.

Thus in truth, the number one or 1 outside the vacuum of space is represented by a cube.

Now, the cube or the number or the element one is composed of three main parts and they are the positive part, the neutral part and the negative part.

But those parts also are composed of 118 other parts.

That is to say, that the positive part also is composed of 118 parts and the neutral part is also composed of 118 parts as the same as the negative part which also is composed of 118 parts.

And when the cube or the number enters into the vacuum of space the cube or the number is compressed into a sphere or into a globe but still with its 118 parts.

Thus, the number one is composed of not only 100 percent but also of another 18 parts or of another 18 percent.

And if we added the parts thus we would have 1.

Now then, the number one or the symbol 1 as also is all of creation is a continuation because the start or the beginning is zero or a point or a neutral or an empty space.

But the number one or the symbol 1 also represents all of existence, the physical part, the neutral part and also the negative part.

Also the number one has the ability of converting itself into its 117 other parts also with their other 118 parts.

The Origin of the Universe / A short version
Forester de Santos

In other words, the number or element one also has the ability of being infinite because also it could renew into a greater number such as the number two.

And that makes it possible the other 117 parts which add to 9 and the number 9 is a symbol of renovation. That first renovation extends the time of the number or of element one.

Thus, if we added the 117 parts which remain to one plus its other two parts, the neutral with its 118 parts, and the negative with its 118 parts, the sum would be of 353.

And if we added 353 thus it would give us 11 or eleven and if we added 11 we would get two, the possibility or the ability of the number or element one if it is renewed.

And once that the number or that element number one has renewed as two, thus the number or element one has become or will continue as double or for much more as two and as double the abundance.

And the very same step or process is with the number or with element number two. If we added all the parts that the number or the element two has, which are the double of one, thus it would give us four or 4.

That is to say, if we added all the parts of the number or element two thus we would get the double.

Now, the number zero or symbol 0 not only represents a reality but also represents a portion of existence as also represents peace or tranquility.

That is the reason that in many cultures the word no is use to attract peace, but the word is misinterpreted. Zero or peace also represents the beginning or start.

The number one represents continuation and also knowledge. And according as to how one was confirmed

The Origin of the Universe / A short version
Forester de Santos

into the world thus one will do and one will hear and one would become as two.

That is to say, the number two represents to hear or to do according to the knowledge given to one as one.

The number three represents unity or united to, thus according to what one has done with the knowledge given to one, one will be united as if three or a third part to him or her that gave knowledge or confirmed one because of one doing or because of one being born alive.

The number four represents adoration or praise because when one is united as if three or as if a third part thus one feels much gladness and much joy and one begins to sing or praise that third part.

The number five represents tests because of the praise which one has done. That praise which one has done because of feeling united with gladness and with joy thus will take one to tests to know if one truly was saying the truth or some confirmation or even a request.

The number six is contention not only because of one being born and being born alive but also the number six is to contend for the consul of him or her that one praised but now has departed from one.

The number seven represents victory which could be granted to one with gladness and with joy and with the feeling of abundance as consul or reward because of the contention of one toward her.

The number eight has much significance. The number eight not only represents all gladness and all joy but also the number eight represents the servant beloved of God or two hearts united as if one.

Also, the number eight reflects or represents eternity or the very grandiose possibility of her.

The Origin of the Universe / A short version
Forester de Santos

The number nine represents or means not only to overcome life or the world but also become as if new or renewed as savior beloved of God and with all the power and with all the authority of the holy heavens and also with all the power of riches and a step much closer to eternity or to immortality.

The number ten represents or means dwelling, in where he that received the very grandiose title of savior beloved of God could be the dwelling beloved of God and God also could be the dwelling beloved of the savior beloved of God.

And if the savior beloved of God accepts being the dwelling of God thus not only God also will be the dwelling beloved of the savior beloved but also the savior beloved will get even closer to eternity or to immortality.

The number eleven represents the double abundance or the double plenitude, where the double abundance or the double plenitude of one is now the double abundance or the double plenitude of the other and the double abundance or the double plenitude of the other now is of one, thus the five portions.

In other words, eleven represents true abundance and because of eleven being true abundance thus eleven renews and eleven also is a step closer to eternity or closer to immortality or closer to the right side of creation or of God.

And finally, the number twelve represents the right side or the eastern side of God or the right side of consciousness or the brain, which truly is an expansion of the conscious mind or a greater identity.

And he that enters to the right side through confirmation of God thus not only is he the right side of God but also he has entered into eternity or into immortality...

The Origin of the Universe / A short version
Forester de Santos

To conclude, just as the number or the element in creation or in the vacuum of space truly is transformed or is converted into another number or into another element thus that way also the conscious being truly can be transformed or converted or enter into the next conscious or the next thinking mode or state.

And that transformation or that power of converting or that entrance into a greater identity truly is done though one's very own mouth…

The Origin of the Universe / A short version
Forester de Santos

Existence

Existence exists because she cannot stop from existing. Existence has no choice but to only exist.

Existence is composed of three portions of lack but which can be seen and existence also is composed of one portion which is but cannot be seen as of yet and which makes for three or even makes for four or for more portions...

Existence is composed of opposites which really are more opposites of themselves than to the opposites to which they are opposites.

And this opposition or contrast which makes possible for existence to attract herself and it appears that existence moves or is in motion because of the very attraction of herself.

And existence is known or is seen more through the little which is seen of her and through the more which is seen of her, less of her is known or less is seen...

And even though existence exists and she attracts herself and she also is eternal knowledge and also she gives knowledge to the vacuum of space or to the universe and thus creating the beginning or the times, existence does not know that she exists even though she also renews herself so that she can continue existing and existing as if always new and as if nothing has happened and will never happen...

The Origin of the Universe / A short version
Forester de Santos

But existence renews herself or is reborn when the vacuum of space becomes empty again or when the knowledge given to the universe is not taken as acknowledgement and she once again gives or once again throws more knowledge into the vacuum of space or into the space of the universe, thus making as if a new beginning and everything which was before as if it never had being or as if it never were...

But that ability which existence has of renewing herself and of continuing for all eternity without knowing end or ending thus we conscious beings also possess.

But we are not obligated to live eternity as neither we are obligated to death or to the end or the ending. Death or the end or the ending comes because it is not done to continue but to continue as if new and to continue forever or to continue for all the times...

Those that do not want eternity thus they just wait for death and death will do for them so that there is no eternity for them...

But those of us that want to continue with life without seeing or without knowing death thus we must do to be reborn and that rebirth or being born again truly is done with knowledge or with acknowledgement.

Thus, we must give knowledge or give acknowledgement to that grandiose part of existence which is but which cannot be seen as of yet but even so it is what gives or it is what grants knowledge not only to the vacuum of space or to the universe but also it is what gives or what grants knowledge to every conscious being that requests it...

And just as one presents oneself to that grandiose part of existence which gives or which grants or which even lends knowledge, thus that grandiose part of existence will be to one.

The Origin of the Universe / A short version
Forester de Santos

In other words, that grandiose or glorious part is the Creating part of existence or the renovating part or the part which gives identity to all existence or which reacts with all the opposites of existence...

Thus, he that gives or that grants or that lends to the Creating part or to the Renovating part of existence the knowledge or the acknowledgement of God or of Creator or of Renovator thus he also will have the knowledge or the acknowledgement of God or of Renovator and God will present to him or come to him or will allow him to draw near with that very same knowledge or acknowledgement...

Existence gives or grants or lends to the vacuum of space or to the universe knowledge in the form of matter or stars or light.

And if that knowledge in the vacuum of space or in the universe is converted into acknowledgement thus the universe will continue as if forever new without ever knowing end or ending nor knowing or remembering beginning...

But if that knowledge given or granted or lend to the vacuum of space or to the universe does not renew or is not converted or is not transformed into acknowledgement, thus that knowledge in the form of matter or stars or light thus will lose her energy and there will only remain and will be space dust no matter how large the piece of space dust and there will only be darkness or there will only be emptiness or vacuum...

But that does not remain like that because now the dust which remained in the vacuum of space or in the vacuum of the universe must be taken out to give or to grant or to lend new knowledge to the vacuum of space or to the universe and that new knowledge will make a new beginning, in

The Origin of the Universe / A short version
Forester de Santos

where there will not be any memory that there ever was a beginning before...

The manner in which existence takes the dust from the vacuum of space or from the universe is with black holes which truly are black vacuum cleaners.

The black vacuums or black holes also suck up any other matter or star which has remained in the vacuum of space or in the space of the universe...

Once there no longer is matter or dust in space thus the black holes also will turn of or will stop from functioning because of lack of energy and they will disperse or they will disintegrate in the vacuum of space.

This emptiness in space, which now has become as if a vacuum or new space because of becoming as before the beginning thus will attract new knowledge in form of matter or stars or light...

And if in that new beginning the same happens which happened in the first, in where there was no renovation or there was no acknowledgement or no rebirth to continue, thus also will have its end or ending even though it may take billions of years...

But that does not have to be as the above because as long as there are conscious beings in the universe, the universe has the very grandiose opportunity of renovating or of rebirth or of receiving acknowledgement so that the universe because of the conscious beings the universe will continue without ever knowing end or ending...

Existence without knowing it renews every time she lends knowledge in the form of matter or in the form of stars or of light to the vacuum of space or to the universe and the universe cannot continue for lack of acknowledgement or for lack of matter or for lack of renovated energy and thus the universe comes to its end or to its ending and thus

The Origin of the Universe / A short version
Forester de Santos

making space for another beginning which will become as if the first beginning and also as if it never had an end or never an ending before...

But if the conscious beings are reborn or revive or take new life or receive acknowledgement through the very same knowledge or acknowledgement which they give or grant or lend, the universe will never ever have end or ending because the universe will continue as if forever new and as if it never had any beginning...

Now then, every conscious being truly has the very grandiose opportunity of being reborn or of reviving or of taking new life or of receiving acknowledgement or rename to be able to continue eternally with life and in complete harmony and in complete abundance...

But if the conscious being does not desire that very grandiose opportunity of living eternally and living in complete harmony and in complete abundance thus that conscious being only has to wait to die and that will he his end or his ending and nothing will become of that conscious being because eternity or immortality is not obligated or is not imposed, even a rock will stop from being or from existing...

Thus, one needs to be alive and conscious to be able to have or can receive immortality in the form of salvation and with her continue alive renewing and renewing also everything else as savior...

Just as knowledge in the form of matter filled with energy or in the form of stars or in the form of light enters the vacuum of space or enters into the universe, thus that same way also thought or knowledge or illumination enters the conscious mind.

That thought or that knowledge or that illumination can take the conscious being to a great state or from one state to

The Origin of the Universe / A short version
Forester de Santos

another state or to a greater identity or from one identity to another identity even though that conscious being really continues with his physical form but every time that that conscious being enters into a greater state of knowledge or into a greater or new identity because of his knowledge, thus the physical form of that conscious being also is refreshed or is seen as if a new form...

But if the thought or if the knowledge or the new identity which enters in that conscious mind of that conscious being is a limited thought or is a limited knowledge or is a limited identity, thus that thought or knowledge or that identity, even though some type of energy or be it negative or be it positive, does not take that conscious being very far or into a greater state of identity and that thought or knowledge or that identity will disperse and the conscious mind becomes once again as if empty.

And if that conscious being nothing does with his conscious mind to have thought or knowledge or identity so that the thought or the knowledge or the identity takes him to a greater state or to a greater identity in where not only his conscious mind will be refreshed or becomes as if a new mind but also his physical form or body also will refresh or even could be cured from certain lacks or faults, such as of that of deafness in one ear or both and also some emotional lack or fault such as loneliness or shyness...

But if the conscious being in the course of his life does not enter into a greater state of thought or of knowledge or of identity, thus the conscious being keeps on dying until he completely dies and his body will discompose until it turns to dust and the conscious being has lost his the very grandiose opportunity of rebirth and of continuing with life as if new in complete or in perfect harmony and also in complete or in perfect abundance, perfect because it will be an abundance which will never ever end...

The Origin of the Universe / A short version
Forester de Santos

Now then, once the universe stops from decomposing or no longer the vacuum of space ever returns to nothing because of the conscious being coming to their maximum state or coming to their maximum identity and that way keeping the universe practically alive, thus there no longer will enter more knowledge in the form of matter or in the form of stars or of light because now that makes it possible the conscious beings because they will be the matter or the stars or the light or the illumination of the universe because of they being illuminated until the maximum or until perfection...

In other words, there will no longer be anymore beginnings nor there will be anymore ends or anymore endings and the universe will become as if there were never beginning because of the universe becoming as if new for all of eternity...

And all the different parts of existence will act or will react as if one as the same the body and the conscious mind of the conscious being will act or will react as if one or as a single part and existence will be one with the conscious being because of the conscious being becoming or as being existence herself and reflecting through his body her glory...

Thus, when one as seed for more came out from the entrails of a man and one entered into the entrails of a woman, one had no memory of that exit or entrance even though one as a seed was in harmony and in abundance in those entrails and one also came out with all gladness and with all joy and also with all feeling of abundance and entered into the entrails of a woman and there also sought for knowledge of life to life receive and in her also enter and once in her also one forgot because once again one entered in harmony while one was transformed or one took the form of life which one did for or for the one which one received

The Origin of the Universe / A short version
Forester de Santos

through the act of one or because of one's movement to find life…

And when one became complete in those new entrails, one humbled and one took the grandiose position of contender and one came out or one entered into the entrails of the world not only as much more but also one came out or one entered for much more.

But in the world one did not remember that one came out from harmony while one keeps completing the form of contender which could take one to not only come to be conscious but also which could take one to rebirth and continue with life as if with a new form.

But if there were no rebirth because of lack of knowledge or because of lack of identity, thus that form not reborn would take one to death and that would be the end or the ending of all of her, life, and also of all of one…

The Origin of the Universe / A short version
Forester de Santos

Knowledge and Chaos

That which is not known or that which is not understood thus that is seen as if chaos even though with some form, but when that is known or when that is truly understood that takes form or takes reform and now is not chaos and now is as if it never lacked form...

Existence truly is all based upon knowledge or be it positive useful knowledge or be it negative or useless knowledge.

There is not an existence completely positive or completely negative because existence exists because of existences being composed of opposite sides which show to be opposites to their very opposites but they are not because they themselves are not a single side or do not exist.

In other words, existence is composed of opposites which show that they are more opposites of themselves than the opposites they are opposites of but that composition is what really makes existence exist...

That is to in truth say, darkness does not shine but darkness is seen. Darkness is not solid but darkness is seen and even though darkness is seen, darkness is lack, lack of light.

The same can be said about space. Space can be seen because of the lack, because of the lack of matter.

The Origin of the Universe / A short version
Forester de Santos

And the same can be said about cold. Cold is felt for lack of heat or lack of warmth.

Thus darkness, space or cold is negative knowledge or is lack of knowledge. But lack does not make herself because lack is made by something which cannot be seen or is not known but that it really is…

In other words, the side that appears to be the right side can be the left side while the south side can be the north side because existence truly is a point of view and the greater the point of view, greater is existence or greater is her grandiose purpose…

Thus, existence truly is composed of four opposite sides which are set up one against the other creating attraction and that attraction also is composed of other opposites, such as positive and negative, which give movement to existence even though existence does not move because of her infinite size, but existence moves through knowledge so that more knowledge she can have as one once moved because of knowledge to have more knowledge so that one could be born and once in the world once again one moved or one did for more knowledge to be able to go on or so that one could continue with life…

Existence is one which truly is composed of infinite numbers or knowledge or be it positive or be it negative which the total of the sum forever truly will add to one, but that total will never be seen by eye but it will be known through understanding and once existence is truly understood by one thus through one existence will be seen because through one existence will reflect…

Thus, existence truly will be seen more because of her lack than her physical presence even though existence is infinite in size and in weight or in useful knowledge.

The Origin of the Universe / A short version
Forester de Santos

But even so without being able to be seen, existence can be truly known and understood because it is true knowledge and it is until the end of all time.

That is to in truth say, existence will never have end because existence is true knowledge and because of being true knowledge, existence makes herself or moves because of knowledge even though existence already is complete and it is eternal...

Knowledge makes possible more knowledge and with that new knowledge existence truly is transformed as if taking a new form.

That in existence, such as matter or the stars or even the atom in the universe, which is not renewed or is not reformed because of lack of knowledge thus that will stop from functioning as such and will become dust and after becoming dust will also stop functioning as dust because it will be separated from all its sides or opposite elements which make possible its existence...

Thus, obviously, the conscious or the living beings can imitate more of existence than existence herself and the very particles that come out from her.

And even though the stars last a lot of time in the vacuum of space or in the universe, the stars with time come to lose their function as stars and begin to turn off.

The stars stop from functioning because they do not have from where to grab from or from where to take out from them very selves to continue functioning as stars because the stars no longer have the matter or the knowledge in themselves to reform or to transform or to renew them very selves...

But the conscious or the living being can truly take out or can truly grab knowledge, which really is matter, from the conscious or the living being him very self just as existence

The Origin of the Universe / A short version
Forester de Santos

her very self or the physical part or the positive part of existence does.

And that truly has to do everything with peace and also with knowledge.

Before the beginning of things there was in the vacuum of space a very profound peace, so profound that it attracted light or matter or knowledge or illumination and thus breaking the very profound peace because the vacuum of space began to compress the light, the matter, the knowledge or what was illumination and that way also creating as if chaos because not only the vacuum of space was no longer the same vacuum or emptiness but also the form of the light or of the matter or of the knowledge or the form of that which was illumination was no longer the same form because the light or the matter or the knowledge or the illumination took another form or reform...

Well then, imagine, if it is that one can imagine, when thought or illumination entered for the very first time into the mind of man and man did not know how to react but he felt as in chaos or in desolation because of lack of peace and because of lack of understanding of that new state of man.

But even so man never again remained as a simple animal because now man had a new identity, the identity of conscious or living being.

And with that new identity man could surpass all the times if man wanted to surpass all the times or to live beyond the times because of the new form or reform of man...

But for man as conscious or as living being can surpass the times, and surpass means taking all the other forms or reforms or states of consciousness or of thought which still remain to be able to continue with life, thus man must do again for his knowledge as once man did for knowledge

The Origin of the Universe / A short version
Forester de Santos

before being born and to be able to be born and thus take the form or the reform or that state of knowledge or of that new identity...

But if man after coming to a new form of knowledge or of identity does not do to come or to get to the next form or reform or new identity, thus man begins to lose his actual form and man dies and man stops from existing as also all the other things in the vacuum of space or in the universe stop from existing because of lack of knowledge or because of lack of new form or reform or even for lack of new identity because everything which stops from existing, even a rock, stops from existing because of lack of new form or because of lack of new identity...

Thus, just as physical existence renews through knowledge because of existence truly being knowledge and that way complete existence taking as if a new form or reform or new identity thus man also for being knowledge through knowledge man also can take as if a new form or reform or a new identity...

But if that new knowledge of man truly begins or is in the conscious mind of man, which gives man identity as man or as conscious or living being and she was granted to man because of some good action or good act of man such as a good feeling of peace or of feeling that in the world as in existence there was more of what man himself could see or could imagine...

In other words, man was illuminated because of his search or because of his good feeling that there was much more in the world as also in existence her very self.

That illumination, which truly is a very grandiose expansion of conscious mind, was granted by the brain or by Him that many of us call God as a form of peace or of consul or as a form of given knowledge or of giving man a new identity of consciousness or living being.

The Origin of the Universe / A short version
Forester de Santos

And once man achieved that new identity of consciousness or of living being, man could share that new knowledge with the rest of mankind so that the rest of mankind also entered into consciousness or that the rest of mankind could be granted the conscious mind.

But that did not remain like that, because now all the sons born after the illumination of man or after man entering into the conscious mind thus also the sons were born conscious or living beings if the sons were born alive...

Thus, when one enters into a new illumination or into a new knowledge or into a greater state of conscious mind or into a new identity, thus one once again must do a new action of confirmation or of request.

But, obviously, this new or this next confirmation or proclamation or even request truly has to be greater or truly superior to the confirmation or the proclamation or even greater or superior to the request of before because one truly will enter for better or for greater into a better or into a greater illumination or into a better or greater knowledge or into a better or greater state of conscious mind or into a new identity which surpasses the former identity...

One truly will know that one has entered into illumination or into a greater state of consciousness when one feels harmony of profound peace, of gladness, of joy, of abundance and one also hears a voice offering or proclaiming such thing as that one will lack nothing or something similar but better or greater than the first illumination or better than the first state of consciousness, in where one truly did not know what was happening to one but even so one became for much more for one truly entering into a greater state of consciousness...

But because of one entering into that new state of consciousness thus one entered as if into chaos because of one not having the peace and not having the knowledge of

The Origin of the Universe / A short version
Forester de Santos

that new state to be able to take the new form or the new identity of one and come out of what seems chaos.

And just as one truly entered into a new conscious state thus also the brain or He who one called God entered into a new state of consciousness or entered into a new heaven in where also peace is lacking and knowledge is lacking to be able to take His greater form or reform or His new identity as Creator and come out with all power and with all authority from what seems to be chaos...

When one was born to the world those that received one comforted one and one rested and one began to adapt to the world while one grew and also one was comforted by those that received or adopted one as son until one dominated or one became accustomed to or one adopted the world and also took the identity with which one was born with.

But when one enters into a greater state of mental consciousness the world can no longer comfort one because the world does not have that peace and does not have that knowledge to complete that new state of knowledge or new identity and the peace that also comes with that knowledge or new state or new identity...

Thus, that is the reason that one has to look up as if into the sky or one has to seek as if in the heavens so that one can truly find or one can truly receive knowledge so that one can enter into the next thinking mode or state of consciousness or into a greater identity or form or reform...

When one was born, one truly was born with the form or reform or with the identity of contender or with that identity of contention and so one triumphed without one knowing and one grew until entering into consciousness or entering into understanding or entering into a new mind.

Thus with that new mind one has to do to continue entering into more consciousness so not to lose the form or the

The Origin of the Universe / A short version
Forester de Santos

reform or the identity of one which truly continues to be of contender, of contender because one truly continues struggling or contending for peace and for knowledge because of one being a conscious being and the conscious being are formed or are reformed or enter into a greater consciousness through knowledge or through acknowledgement but through knowledge or through acknowledgement granted by his very request because what truly grants one's request is the brain or that or He which one calls God or the God of one or even the God of the fathers of one…

Now, with every grant of peace and of knowledge one truly enters into a better or the brain or that which we call our God expands our mind and that expansion or that new state or that new identity truly fills one with gladness and with joy and one also feels much abundance.

But that is not all, because one also hears a voice confirming one's request.

Many times the voice is even heard before one hears the confirmation and one becomes filled with gladness and joy and one also feels with a good sense of abundance or of presence of life even though the abundance or the presence of life is not seen.

But all of that feeling of gladness and of joy and of abundance truly is a very good indicative that there is still more and that the next request will be rewarded also through gladness and through joy and also in all abundance of life or new life…

The Origin of the Universe / A short version
Forester de Santos

Who am I really?

My pen name as a writer is Forester de Santos and I am on a very grandiose crusade of rebirth alive or to be born again with complete gladness and with complete joy and also with complete abundance of God but as much more than God and as much more than Creator.

Now then, one who truly is on a very grandiose crusade cannot follow another or cannot let himself be surrounded by his beloved ones or his fans because he cannot cross over them or he cannot cross over because of them being in the way or because of them blocking the path which is but which cannot be seen until rebirth or until one is born again.

I do not ask to be followed, not because I will not lead, but because I will not look back but I will look to my right and to my left to see who walks with me.

But those that truly decide to follow me will become as me and as me will truly receive or gather true knowledge because my struggle or contention or my very grandiose crusade of rebirth is true, so true in fact that I have become a much better person because of the true faith which I have come to receive through my search and research for the truth.

The Origin of the Universe / A short version
Forester de Santos

And because I have come to have true faith or faith of God, thus I use my true faith as a shield to repel or to reject other beliefs or good sounding lies!

Therefore, to rebirth alive or to be born again while still living here on the very earth which will be as in the very heavens through rebirth!

(0+1)

If you truly enjoyed this simple and humble work, please leave a comment according to your good pleasure and give also a rating but also according to your good pleasure.

Thanks so very much for your time and best of wishes, Forester de Santos.

Thanks for reading my work!

0+1 = peace and knowledge to all mankind!

www.ingramcontent.com/pod-product-compliance
Lightning Source LLC
Chambersburg PA
CBHW070129230526
45472CB00004B/1482